Jack looked around his new room. He did not dislike it. It was just that he missed his old room. He also missed his friends.

Mom and Dad came into Jack's room. "It's time to unpack," said Dad. "Let's untie some of these boxes."

"May I go outside for a little while?" asked Jack. "Then we can unpack."

Mom said, “Don’t stay out too long. I’m going to preheat the stove. Then we’ll have a quick dinner.”

On his way out, Jack passed his little brother, Tim. "Tim does not look unhappy," Jack thought. But Tim was only in preschool. He would not feel the same way as Jack.

"Did you unpack yet?" asked Tim.

"Not yet!" Jack called back. He wished he could be more like Tim. He did not want to displease Mom and Dad.

Jack saw an unknown boy with his dog.

“Hi, I’m Rick, and this is Jumper,” the boy said. “Did you just move in?”

"Yes, I did," Jack answered. "My name is Jack. Do you live close by?"

"I live three houses down," said Rick. "I have to take Jumper home now. He gets unhappy when he's not fed! But maybe we could play tomorrow."

"That would be great!" said Jack. Just then, he saw Mom.

"It's time to have dinner and then unpack!" called Mom.

"You are right!" said Jack, as he went back inside.

The End